SET YOUR COMPASS TRUE

SET YOUR COMPASS TRUE

THE WISDOM OF JOHN, ROBERT & EDWARD KENNEDY

Reflections on Leading an Inspired Life

COMPILED BY SIGNE BERGSTROM

COLLINS DESIGN

An Imprint of HarperCollins Publishers

A nation reveals itself

not only by the men

it produces, but also by

the men it honors,

the men it remembers.

—JOHN F. KENNEDY

CONTENTS

PART I
PUBLIC LIVES AND EVERYDAY VALUES

CONTENTS

CONTENTS

PART II
PRIVATE LIVES AND FAMILY

CONTENTS

PART I

PUBLIC LIVES AND EVERYDAY VALUES

PUBLIC LIVES

What matters about this country cannot be put
into simple slogans; it is a process, a way of doing things
and dealing with people, a way of life.

—ROBERT F. KENNEDY

Yes, we are all Americans. This is what we do.
We reach the moon. We scale the heights. I know it.
I've seen it. I've lived it. And we can do it again.

—EDWARD M. KENNEDY

I am certain that after the dust of centuries has passed
over our cities, we, too, will be remembered
not for victories or defeats in battle or in politics, but for
our contribution to the human spirit.

—JOHN F. KENNEDY

The motto of America…is the Latin phrase "e pluribus unum"—
out of many, one—the whole is greater than the sum of
its parts. The diversity of America is America's greatest strength.

—EDWARD M. KENNEDY

The American, by nature, is optimistic. He is
experimental, an inventor, and a builder who builds best
when called upon to build greatly.

—JOHN F. KENNEDY

We know full well the faults of our democracy—the handicaps of
freedom—the inconvenience of dissent. But I know of no
American who would not rather be a servant in the imperfect house of
freedom, than be a master of all the empires of tyranny.

—ROBERT F. KENNEDY

We hold the view that the people

make the best judgment in the long run.

—JOHN F. KENNEDY

The real strength of our democracy lies not in its institutions,

but in the opportunities it creates for our people.

Our laws, it is said, are "the wise restraints that make us free."

It's the American people who have made this nation

great and who have risen to great challenges in each new era.

—EDWARD M. KENNEDY

America is a nation founded on hope

and the constant quest for a better tomorrow.

—EDWARD M. KENNEDY

Every American ought to have the right to be treated as he would like to be treated, as one would wish to be treated, as one would wish his children be treated.

—JOHN F. KENNEDY

[Martin Luther King Jr.] will never see that dream, but the moment we realize his dream is our dream and work to make it ours, the nation can survive.

—EDWARD M. KENNEDY

Time and time again the American people, facing danger and seemingly insurmountable odds, have mobilized the ingenuity, resourcefulness, strength, and bravery to meet the situation and triumph.

—ROBERT F. KENNEDY

We know that freedom has many dimensions.

It is the right of the man who tills the land to own the land;

the right of the workers to join together to seek

better conditions of labor; the right of businessmen to use

ingenuity and foresight to produce and distribute

without arbitrary interference in a truly competitive economy.

It is the right of government to protect the weak;

it is the right of the weak to find in their courts fair treatment

before the law. It is the right of all our citizens

to engage without fear or constraint in the discussion and

debate of the great issues which confront us all.

We understand this regardless of the extent to which we may

differ in our political views. We know that argument

in the open is one of the sources of our national strength.

—ROBERT F. KENNEDY

I want every American

free to stand up for his rights,

even if sometimes

he has to sit down for them.

—JOHN F. KENNEDY

Ours is a vision of an America

where every person has an equal chance,

where freedom is paramount

and where there is room for the cultivation

of the human spirit.

—EDWARD M. KENNEDY

Everything that makes man's life worthwhile—family, work,

education, a place to rear one's children and

a place to rest one's head—all this depends on decisions of

government; all can be swept away by a government

which does not heed the demands of its people. Therefore,

the essential humanity of men can be protected

and preserved only where government must answer—not

just to the wealthy, not just to those of a particular

religion, or a particular race, but to all its people.

—ROBERT F. KENNEDY

An election is the most important event in our democracy,

and today we reaffirmed that every citizen should have an

equal voice on Election Day. There is nothing more American

than voting, and every American should have the right to vote.

—EDWARD M. KENNEDY

It should be clear that, if one man's rights

are denied, the rights of all are in danger—that if one man

is denied equal protection of the law,

we cannot be sure that we will enjoy freedom of speech

or any other of our fundamental rights.

—ROBERT F. KENNEDY

This nation was founded by men of many nations

and backgrounds. It was founded on

the principle that all men are created equal and that

the rights of every man are diminished

when the rights of one man are threatened.

—JOHN F. KENNEDY

We are no more entitled to oppress

a man for his color

than to shoot a man for his belief.

—EDWARD M. KENNEDY

Health care should be a basic right for all,

not an expensive privilege for the few.

—EDWARD M. KENNEDY

In fact, our values are still our greatest strength.

Despite resistance, setback, and periods

of backlash over the years, our values

have moved us closer to the ideal with which

America began, that all people are created equal.

—EDWARD M. KENNEDY

The glory of justice and the majesty

of the law are created not just by

the Constitution—nor by the courts—nor by

the officers of the law—nor by

the lawyers—but by the men and women

who constitute our society—who are

protectors of the law as they are themselves

protected by the law.

—ROBERT F. KENNEDY

With [John Kennedy's] gift of compassion,

he touched the hearts of peoples

everywhere who believed in this nation's

destiny of freedom and

opportunity and sought it for themselves.

—EDWARD M. KENNEDY

America has a choice. We can continue to be buffeted

by the harsh winds of the global economy.

Or we can think anew, and guide the currents of globalization

with a progressive vision that strengthens the nation

and prepares our people to move more confidently to the

future. In a very real sense, it means revitalizing the

American Dream, so that if people work hard and play by

the rules, they can succeed in life...

—EDWARD M. KENNEDY

Our father used to say that he couldn't have done

any of the things he did, and we couldn't have lived the way we

did, in any other country. He said we owed a great debt

to the government and that is why we ought to work for it.

—ROBERT F. KENNEDY

The challenge of politics and public service

is to discover what is interfering with justice and dignity

for the individual here and now, and then

to decide swiftly upon the appropriate remedies.

—ROBERT F. KENNEDY

My brother Bob doesn't want to be

in government—he promised Dad he'd go straight.

—JOHN F. KENNEDY

People like myself can't go around

making nice speeches all the time. We can't

just keep raising expectations.

We have to do some damn hard work, too.

—ROBERT F. KENNEDY

Mothers all want their sons to grow up to be president,

but they don't want them to become politicians in the process.

—JOHN F. KENNEDY

President Kennedy loved to tell the story of the famous Gallup Poll of

his day, which found that parents wanted their children

to be president—as long as they didn't have to be involved in politics.

It was a line that always got a laugh—but to Jack it was hardly a

laughing matter. He loved and lived by the famous words of John

Bunyan in *Pilgrim's Progress*, that: "Public life is the crown

of a career, and to young persons it is the worthiest ambition. Politics

is still the greatest and most honorable adventure."...Yes, politics is

still the greatest and most honorable adventure. And as long as you

carry within you the faith that those words are true, they'll remain

true, no matter how that faith is challenged by day-to-day events.

—EDWARD M. KENNEDY

I like politics. It's an honorable adventure.

—ROBERT F. KENNEDY

Politics is a jungle.

—JOHN F. KENNEDY

In politics, I think, as in life, a lot of it is also being at the right place

at the right time…I mean that's certainly happened to me.

—EDWARD M. KENNEDY

Those of you who regard my profession of political life

with some disdain should remember that it made it

possible for me to move from being an obscure lieutenant in the

United States Navy to commander in chief in

fourteen years with very little technical competence.

—JOHN F. KENNEDY

On his brother Edward:

The best politician in the family.

—JOHN F. KENNEDY

The pursuit of the presidency

is not my life, public service is.

—EDWARD M. KENNEDY

My experience in government

is that when things are

noncontroversial and beautifully coordinated,

there is not much going on.

—JOHN F. KENNEDY

I grew up knowing that we had many advantages

other people didn't have. It was understood

that with such advantages went responsibilities.

My parents gave us all the feeling that they'd do

anything they could to help us meet those responsibilities,

which we understood were for some kind of public

service. I guess we all knew we could goof off, but there was

a strong challenge to make something of ourselves.

—EDWARD M. KENNEDY

Sometimes we all have to do things we don't want to do.

—JOHN F. KENNEDY

My fellow citizens of the world: ask not what America will do

for you, but what together we can do for the freedom of man.

—JOHN F. KENNEDY

Well, I wanted to be a newspaperman, but

my brother was killed in the war,

and my dad thinks I'm best fitted to carry on for him.

You want to be a lawyer, and I want to be

a newspaperman, but if we are going to change things

the way they should be changed,

we have to do things we don't want to do.

—JOHN F. KENNEDY

If I died, my brother Bob

would want to be senator. And if anything

happened to him,

my brother Ted would run for us.

—JOHN F. KENNEDY

Like my three brothers before me,

I pick up a fallen standard.

Sustained by their memory of our priceless

years together I shall try to carry

forward that special commitment to justice,

to excellence, to courage

that distinguished their lives.

—EDWARD M. KENNEDY

A man does what he must—in spite

of personal consequences,

in spite of obstacles and dangers and

pressures—and that is

the basis of all human morality.

—JOHN F. KENNEDY

In the long history of the world, only a few

generations have been granted

the role of defending freedom in its hour

of maximum danger. I do not shrink

from this responsibility—I welcome it. I do not

believe that any of us would exchange

places with any other people or any other generation.

The energy, the faith, the devotion which

we bring to this endeavor will light our country

and all who serve it—and the glow

from that fire can truly light the world.

—JOHN F. KENNEDY

If not us, who? If not now, when?

—ROBERT F. KENNEDY

A young man who does not have

what it takes to perform military service is not likely

to have what it takes to make a living.

—JOHN F. KENNEDY

You must wonder when it is all going to end and

when we can come back home. Well, it isn't going to end....

We have to stay at it. We must not be fatigued.

—JOHN F. KENNEDY

No one expects that our life will be easy....

History will not permit it...[But] we...will continue

to do, as we have done in our past, our duty.

—JOHN F. KENNEDY

I agree that our standard that we hold

up to the rest of the world

might be higher and might be different,

and therefore we have

a greater responsibility to adhere to it.

—EDWARD M. KENNEDY

The responsibility of our time

is nothing less than to lead

a revolution—a revolution which will be

peaceful if we are wise enough;

human if we care enough; successful if we are

fortunate enough—but a revolution

which will come whether we will it or not.

—ROBERT F. KENNEDY

For of those to whom much is given, much is required.

And when at some future date the high court

of history sits in judgment on each one of us, recording

whether in our brief span of service we fulfilled

our responsibilities to the state, our success or failure,

in whatever office we may hold, will be measured

by the answers to four questions: First, were we truly men

of courage—with the courage to stand up

to one's enemies—and the courage to stand up, when

necessary, to one's own associates—the courage

to resist public pressure, as well as private greed?

—JOHN F. KENNEDY

Great crises make great men.

—JOHN F. KENNEDY

Throughout our history, when leaders have provided hope
and have unified the country in a common vision, Americans have risen
to the challenge. That is still part of the American spirit today.
Even as we struggle with these constant challenges to our constitutional
system and our ideals, there are still times when we in the political arena
are able to step up to the plate, try our best, and get something right.

—EDWARD M. KENNEDY

Our problems are man-made, therefore they may be solved
by man. No problem of human destiny is beyond human beings.

—JOHN F. KENNEDY

The more I discover about other people's personal lives
the more I see that every household has…problems, in one form
or another. People do the best they can.

—EDWARD M. KENNEDY

The sharpest criticism often goes hand in hand with the
deepest idealism and love of country.

—ROBERT F. KENNEDY

We love our country for what it can be and for the justice
it stands for and what we're going to mean to the next
generation. It is not just the land, it is not just the mountains,
it is what this country stands for.

—ROBERT F. KENNEDY

Our country demands a great deal from us, and
we rightly demand a great deal from our leaders. America is a
compact, a bargain, a contract. It says that all of us
are connected. Our fates are intertwined. Fifty states, one
nation. Our Constitution binds us together.

—EDWARD M. KENNEDY

The cost of freedom is always high—and
Americans have always paid it. And one path we shall
never choose, and that is the path of surrender or
submission. Our goal is not the victory of might, but
the vindication of right—not peace at the expense
of freedom, but both peace and freedom, here in this
hemisphere, and, we hope, around the world.

—JOHN F. KENNEDY

What really counts is not the immediate act
of courage or valor, but those who bear the struggle day
in and day out—not the sunshine patriots but
those who are willing to stand for a long period of time.

—JOHN F. KENNEDY

Yet the gross national product does not allow

for the health of our children, the quality of their education,

or the joy of their play. It does not include

the beauty of our poetry or the strength of our marriages,

the intelligence of our public debate or

the integrity of our public officials. It measures neither

our wit nor our courage, neither our wisdom nor

our learning, neither our compassion nor our devotion to

our country; it measures everything, in short,

except that which makes life worthwhile. And it can tell us

everything about America except

why we are proud that we are Americans.

—ROBERT F. KENNEDY

Freedom is indivisible,

and when one man

is enslaved, all are not free....

—JOHN F. KENNEDY

Conformity is

that jailer of freedom and

the enemy of growth.

—JOHN F. KENNEDY

The unity of freedom

has never relied

on uniformity of opinion.

—JOHN F. KENNEDY

The citizens of our democracy have a fundamental right

to debate and even doubt the wisdom of a president's policies.

And the citizens of our democracy have a sacred obligation

to sound the alarm and shed light on the policies of an

administration that is leading this country to a perilous place.

—EDWARD M. KENNEDY

If we cannot now end our differences,

at least we can help make the world safe for diversity.

—JOHN F. KENNEDY

Only the strength and progress and peaceful change that come from

independent judgment and individual ideas—and even from

the unorthodox and the eccentric—can enable us to surpass that foreign

ideology that fears free thought more than it fears hydrogen bombs.

—JOHN F. KENNEDY

For in the final analysis, our most

basic common link is

that we all inhabit this small planet,

we all breathe the same air,

we all cherish our children's futures,

and we are all mortal.

—JOHN F. KENNEDY

The supreme reality of

our time is the vulnerability of our planet.

—JOHN F. KENNEDY

I think we're all very entirely different people,

but there were common kinds

of interests and common concerns.

—EDWARD M. KENNEDY

For as the Apostle Paul wrote long ago in Romans:

"If it be possible, as much as it lives in you,

live peaceable with all men." I believe it is possible;

the choice lies within us; as fellow citizens,

let us live peaceable with each other; as fellow human

beings, let us strive to live peaceably with men

and women everywhere. Let that be our purpose and

our prayer, yours and mine—for ourselves,

for our country, and for all the world.

—EDWARD M. KENNEDY

World peace, like community peace, does not

require that each man love his neighbor—it requires only

that they live together with mutual tolerance....

—JOHN F. KENNEDY

Equality and mutual respect are the twin pillars of peace.

—EDWARD M. KENNEDY

It is one of the prices of democracy that
there is sometimes a difference of viewpoint. But sometimes
we benefit from these differences.

—ROBERT F. KENNEDY

If [John Kennedy's] life and death had a meaning,
it was that we should not hate but love one another, we should
use our powers not to create conditions of oppression that
lead to violence, but conditions of freedom that lead to peace.

—EDWARD M. KENNEDY

The war against hunger is truly man's war of liberation.

—JOHN F. KENNEDY

Whenever men take the law into their own hands,

the loser is the law. And when the law loses, freedom languishes.

—ROBERT F. KENNEDY

Let both sides seek to invoke the wonders of science

instead of its terrors. Together let us explore

the stars, conquer the deserts, eradicate disease, tap the ocean

depths, and encourage the arts and commerce.

—JOHN F. KENNEDY

All of us have in our veins the exact same percentage of salt in

our blood that exists in the ocean, and, therefore, we have

salt in our blood, in our sweat, in our tears. We are tied to the

ocean. And when we go back to the sea—whether it is to sail or

to watch it—we are going back from whence we came.

—JOHN F. KENNEDY

On the first day of school, the math teacher

made a small speech to the class in which he said

that two great things had happened to him:

one that Rommel was surrounded in Egypt and second

that Kennedy had passed a math test.

—ROBERT F. KENNEDY

You know nothing for sure…

except the fact

that you know nothing for sure.

—JOHN F. KENNEDY

Leadership and learning

are indispensable to each other.

—JOHN F. KENNEDY

I was fourteen when Jack ran for Congress in 1946,

but I remember what he told me

shortly after he took office. He was taking me

around Washington, pointing out

the different landmarks....The thing that is seared

in my memory and that has influenced

the rest of my life is what my brother said to me

at the end of our day of touring.

"It's good that you're interested in seeing

these buildings, Teddy. But I hope you

also take interest in what goes on inside them."

—EDWARD M. KENNEDY

We celebrate the past to awaken the future.

—JOHN F. KENNEDY

America's founders recognized

educational opportunity as an enduring truth.

A strong education encourages

good citizenship. It strengthens our economy

and our national defense, and enables new

generations of Americans to fulfill their dreams.

It is the true path to opportunity.

—EDWARD M. KENNEDY

The ignorance of one voter

in a democracy

impairs the security of all.

—JOHN F. KENNEDY

Speaking about problems facing the nation on the Merv Griffin Show:

I think another great problem in our country…

particularly among young people…is that everything's become

so impersonal. Everything's so large, so big, and you feel

that you don't play any role anymore, that you can't

effect things, can't effect what the government does….

It spends a great deal of money, but you are sort of a small cog.

You have very little say in how a university is run or what

you're being taught…. You're just a number in a school.

—ROBERT F. KENNEDY

Someone was kind enough, though I don't know

whether he meant it kindly, to say the other night that in my campaign

in California I sounded like a Truman with a Harvard accent.

—JOHN F. KENNEDY

I suspect there may always be arguments about what
constitutes a higher education, but wise men through the ages
have at least been able to agree on its purpose.
Its purpose is not only to discipline and instruct, but above all
to free the mind—to free it from the darkness,
the narrowness, the groundless fears and self-defeating
passions of ignorance. You may sometimes regret it,
for a free mind insists on seeking out reality, and reality is
often a far more painful matter than the soft
and comfortable illusions of the intellectually poor.

—ROBERT F. KENNEDY

Your role as university men, whatever your calling, will be
to increase each new generation's grasp on its duties.

—JOHN F. KENNEDY

A president of Harvard is reported to have said that

the reason universities are such great storehouses of learning

is that every entering student brings a little knowledge in—

and no graduating student ever takes any knowledge out.

—EDWARD M. KENNEDY

Upon receiving an honorary degree from Yale:

It might be said now that I have the best of both worlds,

a Harvard education and a Yale degree.

—JOHN F. KENNEDY

To further the appreciation of culture among all the people,

to increase respect for the creative individual, to widen

participation by all the processes and fulfillments of art—this

is one of the fascinating challenges of these days.

—JOHN F. KENNEDY

If sometimes our great artists have been

the most critical of our society, it is because their sensitivity

and their concern for justice,

which must motivate any true artist, makes him aware that

our nation falls short of its highest potential.

—JOHN F. KENNEDY

If art is to nourish the roots of our culture,

society must set the artist free

to follow his vision wherever it takes him.

—JOHN F. KENNEDY

If more politicians knew poetry,

and more poets

knew politics, I am convinced

the world would be

a little better place to live.

—JOHN F. KENNEDY

We believe that an artist,

in order to be true

to himself and his work,

must be a free man.

—JOHN F. KENNEDY

Asked if the arts can help heal:

No question. I mean, it reaches the deepest emotions,

the highest aspiration. I really believe

it's the highest form of human achievement.

—EDWARD M. KENNEDY

I look forward to an America

which will reward achievement in the arts as we reward

achievement in business or statecraft.

—JOHN F. KENNEDY

If you read history and understand the movement of either

Western political thought or even Eastern…

you'll find that the greatest days of art were also the greatest

days of sensible and responsible leadership in public life.

—EDWARD M. KENNEDY

The life of the arts, far from

being an interruption,

a distraction, in the life of a nation,

is very close to the center

of a nation's purpose—and is a test of the

quality of a nation's civilization.

—JOHN F. KENNEDY

In free society art is not a weapon....

Artists are not engineers of the soul.

—JOHN F. KENNEDY

I look forward to an America

which will not be afraid of grace and beauty.

—JOHN F. KENNEDY

We don't want to be

like the leader in the French Revolution

who said, "There go my people.

I must find out where they are going

so I can lead them."

—JOHN F. KENNEDY

Nations around the world

look to us for leadership not merely

by strength of arms,

but by the strength of our convictions.

—ROBERT F. KENNEDY

War and violence, hunger and poverty, injustice and abuse
of power are as old as the human race. But they are not an
unchangeable result of our DNA. We know from our history…
that the right leadership can summon, as President Lincoln
said, "the better angels of our nature" and inspire us
to meet our challenges and make the world a better place.

—EDWARD M. KENNEDY

In the decade that lies ahead—in the challenging,
revolutionary sixties—the American presidency will demand
more than ringing manifestos issued from the rear of battle.
It will demand that the president place himself in the very
thick of the fight, that he care passionately about the fate
of the people he leads, that he be willing to serve them, at the
risk of incurring their momentary displeasure.

—JOHN F. KENNEDY

Dante once said that the hottest places in hell

are reserved for those

who in a period of moral crisis maintain their neutrality.

—JOHN F. KENNEDY

And the task of leadership, the first task of concerned people,

is not to condemn or castigate or deplore.

It is to search out the reason for disillusionment and alienation,

the rationale of protest and dissent—perhaps,

indeed, to learn from it.

—ROBERT F. KENNEDY

We, the people, are the boss, and we will get

the kind of political leadership,

be it good or bad, that we demand and deserve.

—JOHN F. KENNEDY

I suppose anybody in politics

would like to be president.... At least you have

an opportunity to do something about all

the problems, which I would be concerned about...

as a father or as a citizen.

—JOHN F. KENNEDY

I do not run for the office

of the presidency after fourteen years

in the Congress with any expectation

that it is an empty or an easy job.

I run for the presidency of the United States

because it is the center of action.

—JOHN F. KENNEDY

Do you realize the responsibility I carry?

I'm the only person standing

between Nixon and the White House.

—JOHN F. KENNEDY

I want to express my great appreciation

to all of you for your kindness

in coming out and giving us a warm Hoosier welcome.

I understand that this town suffered

a misfortune this morning when the bank was robbed.

I am confident that the *Indianapolis Star*

will say "Democrats Arrive and Bank Robbed."

But we don't believe that.

—JOHN F. KENNEDY

If I am elected, I don't want to

wake up on the morning of November 9

and have to ask myself,

"What in the world do I do now?"

—JOHN F. KENNEDY

On the possibility of running for a second term:

It will be tough.

But then everything is tough.

—JOHN F. KENNEDY

I personally have lived through ten presidential campaigns,

but I must say the eleventh

makes me feel like I lived through twenty-five.

—JOHN F. KENNEDY

On his ambivalence about running for president:

Yeah, I think I'll run.

Maybe I'll run. Yeah, I think I'm going to run.

—ROBERT F. KENNEDY

After he'd declared that he was running for president:

I'm sleeping well for the first time

in months. I don't know what's going to happen,

but at least I'm at peace with myself.

—ROBERT F. KENNEDY

We are a great country,

an unselfish country and a compassionate country.

I intend to make that my basis for running.

—ROBERT F. KENNEDY

I am running because this country

is on such a perilous course and because I have

such strong feelings about what must be done,

and I feel I am obliged to do all that I can.

—ROBERT F. KENNEDY

I can accept the fact that I may not be

nominated now. If that happens,

I will just go back to the Senate, and say what

I believe, and not try in 1972.

Somebody has to speak up for the Negroes and

Indians and Mexicans and poor whites.

Maybe that's what I do best.

Maybe my personality just isn't built for this.

—ROBERT F. KENNEDY

I finally realized I was

in a horse race, and

I wasn't necessarily the favorite.

—ROBERT F. KENNEDY

Don't ever run

for p-p-p-president. It's very tiring.

—ROBERT F. KENNEDY

There are two roads

to the [presidential] nomination,

one is to seek commitments

through discussions with political leaders,

the other is to go to the people.

—ROBERT F. KENNEDY

After winning the Nebraska primary:

I've got to go because I have thousands of fans

waiting for me—I hope.

—ROBERT F. KENNEDY

On his decision to run for president:

Ted thinks I'm a little nutty for doing this,

but he's an entirely different person.

—ROBERT F. KENNEDY

I think the only person

who can find answers for the United States

is probably God.

And unfortunately he isn't running.

—ROBERT F. KENNEDY

PREVIOUS PAGE: The Kennedys in Hyannis Port, Massachusetts, ca. 1930s. Clockwise from center: Joseph P. Kennedy Sr. (seated), Edward, John, Joseph Jr., and Robert.

ABOVE: John, Robert, and Edward in Hyannis Port, Thanksgiving, 1948. They were 31, 23, and 17 years old respectively at the time.

RIGHT: The Kennedy clan, Thanksgiving, 1948. From left: John, Jean, Rose, Joseph Sr., Patricia, Robert, Eunice, and Edward.

LEFT: John F. and Robert F. Kennedy playing football, 1957.

BELOW: John F. Kennedy throws stones along the surf in Hyannis Port, 1953.

RIGHT: The Kennedy brothers, Washington, D.C., 1958.

NEXT PAGE: Edward M. and Robert F. Kennedy attend a Senate hearing, 1967.

Well, if I were to run, which

I don't intend to, I would hope to win.

—EDWARD M. KENNEDY

After his unsuccessful run for the presidency,

Edward Kennedy told the delegates

at the 1980 Democratic Convention in New York:

Well, things worked out a little different

from the way I thought, but let me tell you,

I still love New York.

When we got into office,

the thing that surprised me most was

to find that things were just

as bad as we'd been saying they were.

—JOHN F. KENNEDY

Sure, it's a big job;

but I don't know anyone who can do it

better than I can.

—JOHN F. KENNEDY

To a group of students who were working in

Washington for the summer:

Sometimes I wish I just had a summer job here.

—JOHN F. KENNEDY

The day before my inauguration,

President Eisenhower told me, "You'll find

that no easy problems ever

come to the president of the United States.

If they are easy to solve, somebody else

has solved them." I found that hard to believe,

but now I know it is true.

—JOHN F. KENNEDY

I have a nice home,

the office is close by and the pay is good.

—JOHN F. KENNEDY

Remark made during the Cuban Missile Crisis:

I guess this is the week I earn my salary.

—JOHN F. KENNEDY

There is no experience

you can get that can possibly prepare you

adequately for the presidency.

—JOHN F. KENNEDY

Only the president himself

can know what his real pressures and

his real alternatives are.

—JOHN F. KENNEDY

What is prestige? Is it the shadow of

power or the substance of power? We are going to

work on the substance of power.

—JOHN F. KENNEDY

This isn't the way they told me it was when

I first decided to run for the presidency.

After reading about the schedules of the president,

I thought we all stayed in bed until ten

or eleven and then got out and drove around.

—JOHN F. KENNEDY

I want to express my appreciation to the governor.

Every time he introduces me as the potentially

greatest president in the history of the United States, I always

think perhaps he is overstating it one or two degrees.

George Washington wasn't a bad president and

I do want to say a word for Thomas Jefferson.

But, otherwise, I will accept the compliment.

—JOHN F. KENNEDY

When asked what his favorite song was:

I think "Hail to the Chief" has a nice ring to it.

—JOHN F. KENNEDY

I am only an ivory-tower president.

—JOHN F. KENNEDY

There is no question about it—in the next

forty years a Negro can achieve

the same position that my brother has.

—ROBERT F. KENNEDY

[John Kennedy] made

America feel young again.

—ROBERT F. KENNEDY

John Kennedy was not only

president of one nation; he was president

of young people around the world.

—ROBERT F. KENNEDY

About John Kennedy's presidency:

It was all so brief.

The thousand days are like an evening gone.

—EDWARD M. KENNEDY

Thomas Jefferson once wrote

that a little rebellion

now and then is a good thing. But if I'm

elected president...don't try it.

—ROBERT F. KENNEDY

Frankly, I don't mind

not being president. I just mind

that someone else is.

—EDWARD M. KENNEDY

To exclude from positions of trust

and command all those below the age of forty-four would

have kept Jefferson from writing

the Declaration of Independence, Washington from

commanding the Continental Army,

Madison from fathering the Constitution, Hamilton

from serving as secretary of the treasury, Clay from

being elected Speaker of the House, and

Christopher Columbus from discovering America.

—JOHN F. KENNEDY

Nixon is about as far advanced as I was ten years ago.

—JOHN F. KENNEDY

Ladies and gentlemen, the outstanding

news story of this week was not the events of the

United Nations or even the presidential campaign.

It was a story coming out of my own city of Boston that

Ted Williams of the Boston Red Sox had retired

from baseball. It seems that at forty-two he was too old.

It shows that perhaps experience isn't enough.

—JOHN F. KENNEDY

It has recently been suggested

that whether I serve one or two terms in the presidency,

I will find myself at the end of that period

at what might be called the awkward age, too old to begin

a new career and too young to write my memoirs.

—JOHN F. KENNEDY

Bobby and I smile sardonically.

Teddy will learn how to smile sardonically in two or three

years, but he doesn't know how, yet.

—JOHN F. KENNEDY

When I was young,

I thought the most important thing was being young

and having energy, but now

I say there's nothing like experience.

—EDWARD M. KENNEDY

With the continuing support

of the people of Massachusetts, I intend to stay in this job

until I get the hang of it.

—EDWARD M. KENNEDY

EVERYDAY VALUES

The problems of the world cannot possibly

be solved by skeptics or cynics whose horizons

are limited by the obvious realities.

We need men who can dream of things that

never were and ask why not.

—JOHN F. KENNEDY

Sometimes a party must sail against the wind.

We cannot afford to drift or lie at anchor. We cannot heed

the call of those who say it is time to furl the sail.

—EDWARD M. KENNEDY

We should not let our fears

hold us back from pursuing our hopes.

—JOHN F. KENNEDY

All things are to be examined

and called into question—there are

no limits set to thought.

—ROBERT F. KENNEDY

A man may die,

nations may rise and fall,

but an idea lives on.

—JOHN F. KENNEDY

Every president must endure

the gap between what he would like

and what is possible.

—JOHN F. KENNEDY

A politician is a dream merchant.

But he must back up the dream.

—JOHN F. KENNEDY

The future does not belong to those

who are content with today,

apathetic towards common problems

and their fellow man alike, timid and fearful

in the face of new ideas and bold projects....

It will belong to those who see

that wisdom can only emerge from the clash

of contending views, the passionate expression

of deep and hostile beliefs. Plato said:

"A life without criticism is not worth living."

—ROBERT F. KENNEDY

But I believe the times demand new invention,

innovation, imagination, decision.

I am asking each of you to be pioneers on that New Frontier.

My call is to the young in heart, regardless of age—to

all who respond to the Scriptural call: "Be strong and of a good

courage; be not afraid, neither be thou dismayed."

—JOHN F. KENNEDY

The youthfulness I speak of is not a time of life,

but a state of mind, a temper of the will,

a quality of the imagination, a predominance of courage

over timidity, of the appetite for adventure over the love

of ease…. It does not accept the failures

of today as a reason for the cruelties of tomorrow.

—ROBERT F. KENNEDY

There are risks and costs to a program of action.

But they are far less than the long-range risks and costs

of comfortable inaction.

—JOHN F. KENNEDY

Only those who dare to fail greatly can ever achieve greatly.

—ROBERT F. KENNEDY

Explaining the traits needed in an attorney general:

I don't want somebody who is going to be

faint-hearted. I want somebody who is going to be strong,

who will join with me in taking whatever risks...and

who would deal with the problem honestly.

—JOHN F. KENNEDY

History is a relentless master.

It has no present, only the past rushing into the future.

To try to hold fast is to be swept aside.

—JOHN F. KENNEDY

The complacent, the self-indulgent,

the soft societies are about to be swept away with

the debris of history.

—JOHN F. KENNEDY

The State Department is a bowl of jelly.

It's got all those people over there

who are constantly smiling. I think we need to

smile less and be tougher.

—JOHN F. KENNEDY

Courage is the virtue that President Kennedy

most admired. He sought out those people

who had demonstrated in some way, whether it was on

a battlefield or a baseball diamond, in a speech

or fighting for a cause, that they had courage, that they

would stand up, that they could be counted on.

—ROBERT F. KENNEDY

[John Kennedy] had courage

which permitted him to bear sharp pain

almost every day and which made him

respond to the pain of others. He rejected the

cold affliction of indifference and

the comfortable erosion of concern.

—EDWARD M. KENNEDY

The courage of life is often a less

dramatic spectacle than

the courage of a final moment;

but it is no less a magnificent

mixture of triumph and tragedy.

—JOHN F. KENNEDY

If we are strong,

our strength will speak for itself.

If we are weak,

words will be no help.

—JOHN F. KENNEDY

There is no safety in hiding.

—EDWARD M. KENNEDY

In whatever area in life one may meet

the challenges of courage,

whatever may be the sacrifices he faces if he follows

his conscience—the loss of friends,

his fortune, his contentment, even the esteem of

his fellow man—each man must decide

for himself the course he will follow. The stories

of past courage can define that ingredient—

they can teach, they can offer hope,

they can provide inspiration. But they

cannot supply courage itself. For this, each man must

look into his own soul.

—JOHN F. KENNEDY

War will exist until

that distant day when the conscientious objector

enjoys the same reputation

and prestige that the warrior does today.

—JOHN F. KENNEDY

Everyone admires courage

and the greenest garlands are for those who possess it.

—JOHN F. KENNEDY

Neither smiles nor frowns,

neither good intentions nor harsh words,

are a substitute for strength.

—JOHN F. KENNEDY

This is a great nation and a strong people.

Any who seek to comfort

rather than speak plainly, reassure rather than instruct,

promise satisfaction rather than

reveal frustration—they deny that greatness and drain

that strength. For today as it was

in the beginning, it is the truth that makes us free.

—ROBERT F. KENNEDY

The great enemy of the truth is very often not the lie—

deliberate, contrived, and dishonest—but the myth—

persistent, persuasive, and unrealistic. Too often we hold fast

to the clichés of our forebears. We subject all facts

to a prefabricated set of interpretations. We enjoy the comfort

of opinion without the discomfort of thought.

—JOHN F. KENNEDY

Today's heresy

has a disconcerting potential for

becoming tomorrow's truth.

—EDWARD M. KENNEDY

Truth is a tyrant—the only tyrant

to whom we can give our allegiance. The service

of truth is a matter of heroism.

—JOHN F. KENNEDY

The great challenge to all Americans—

indeed to all free men and women—is to

maintain loyalty to truth.

—ROBERT F. KENNEDY

Fear not the path of truth

for the lack of people walking on it.

—ROBERT F. KENNEDY

On his brother Robert:

There really is no person with whom

I have been intimately connected

over the years. I need to know that when

problems arise I'm going to have

somebody who's going to tell me

the unvarnished truth, no matter what.

—JOHN F. KENNEDY

Let [the television networks] show the sound,

the feel, the hopelessness, and what it's like to think you'll

never get out. Show a black teenager, told by

some radio jingle to stay in school, looking at his older

brother, who stayed in school and is out of a job.

Show the Mafia pushing narcotics; put a *Candid Camera* team

in a ghetto school and watch what a rotten system of

education it really is. Film a mother staying up all night to

keep threats from her baby.... Then I'd ask people

to watch it and experience what it means to live in the most

affluent society in history—without hope.

—ROBERT F. KENNEDY

We must never forget that art

is not a form of propaganda; it is a form of truth.

—JOHN F. KENNEDY

In each of us, there is a private hope

and dream which, fulfilled, can be translated

into benefit for everyone.

—JOHN F. KENNEDY

President Kennedy believed very strongly,

as he often said, that each of us

can make a difference and all of us should try....

The leaders of his generation were

motivated by...a sense that all things are possible

for those who dare to dream, who are

bold enough to believe that together we can leave

this world better than we found it.

—EDWARD M. KENNEDY

All around us, we see the politics of hope being replaced

by the politics of fear, and we ask,

how we can change things? There is a way, and

it's called committing ourselves to

the public square—becoming involved in our communities—

speaking out for what's right, what's fair,

what's just. Time and time again throughout our history,

we've learned that we have the capacity

to find solutions to our problems. All we have to do is try.

—EDWARD M. KENNEDY

It is the glory and the greatness of our tradition

to speak for those who have no voice,

to remember those who are forgotten.

—JOHN F. KENNEDY

[John Kennedy's] appeal to young people

to make a difference in terms of their community—

whether it was elective office or others—

was really, I think, probably the most enduring

and lasting part of his legacy.

—EDWARD M. KENNEDY

If there is a lesson from [John Kennedy's] life

and from his death,

it is that in this world of ours

none of us can afford to be lookers-on,

the critics standing on the sidelines.

—ROBERT F. KENNEDY

Each time a man stands up for an ideal,

or acts to improve the lot of others,

or strikes out against injustice, he sends forth a tiny ripple

of hope, and crossing each other

from a million different centers of energy and

daring those ripples build a current which can sweep down

the mightiest walls of oppression and resistance.

—ROBERT F. KENNEDY

It is not given to us to right every wrong,

to make perfect all the imperfections of the world.

But neither is it given to us to sit content in our storehouses—

dieting while others starve, buying eight million

new cars a year while most of the world goes without shoes.

—ROBERT F. KENNEDY

All great questions must be raised

by great voices,

and the greatest voice is the voice

of the people—speaking out—in prose,

or painting or poetry or music;

speaking out—in homes and halls, streets and farms,

courts and cafes—let that voice

speak and the stillness you hear will be

the gratitude of mankind.

—ROBERT F. KENNEDY

Few will have the greatness to bend history itself,

but each of us can work to change

a small portion of events, and in the total of all those acts

will be written the history of this generation.

—ROBERT F. KENNEDY

I think you have one time around

and I don't know what's going to be in existence

in six months or a year. I think that

there are all kinds of these problems [to] which

you are here on earth to make some

contribution of some kind. I think if one feels involved

one should try to do something.

—ROBERT F. KENNEDY

[Robert Kennedy] need not be idealized

or enlarged in death beyond what he was in life.

He should be remembered as a good and decent man

who saw wrong and tried to right it, saw suffering and tried

to heal it, saw war and tried to stop it.

—EDWARD M. KENNEDY

In 1968, at a time of unconscionable violence

in America, my brother Robert Kennedy

spoke of the dream of peace and an end to conflict,

in words that summon us all to action now:

"It is up to those who are here—fellow citizens and

public officials—to carry out that dream,

to try to end the divisions that exist so deeply in

our country and to remove

the stain of bloodshed from our land."

—EDWARD M. KENNEDY

If there was one great meaning

to Robert Kennedy's campaign...it was that voting

every four years was not enough

to make a citizen—and not enough to satisfy a man.

Rather, each of us must...do his individual part

to end the suffering, feed the hungry,

heal the sick and strengthen and renew the national spirit.

—EDWARD M. KENNEDY

None of us has all the answers

to the challenges that confront us today, but

we each have the ability to make a difference.

All we need is the will.

—EDWARD M. KENNEDY

A fair prosperity and a just society

are within our vision

and our grasp, and we do not have every answer.

There are questions

not yet asked, waiting for us in the recesses

of the future, but of this much

we can be certain because it is the lesson

of all our history:

Together a president and

the people can make a difference.

—EDWARD M. KENNEDY

There is no sense in

raising hell, and then not being successful.

—JOHN F. KENNEDY

You say to yourself,

"I wonder if I can do it," and then later

you might say,

"I think I can do it," and you try

and you succeed,

and it's a wonderful thing.

—EDWARD M. KENNEDY

Wishing it, predicting it,

even asking for it, will not make it so.

—JOHN F. KENNEDY

If we fail to dare, if we do not try,

the next generation will harvest the fruit of our indifference;

a world we did not want—a world we did not choose—but

a world we could have made better.... And we shall be left

only with the hollow apology of T. S. Eliot:

"That is not what I meant at all. That is not it, at all."

—ROBERT F. KENNEDY

As every past generation has had

to disenthrall itself from an inheritance of truisms and

stereotypes, so in our time we must move on

from the reassuring repetition of stale phrases to a new,

difficult, but essential confrontation with reality.

—JOHN F. KENNEDY

One-fifth of the people

are against everything all the time.

—ROBERT F. KENNEDY

I feel that death is the end of a lot of things.

I just hope the Lord gives me

the time to get all these things done.

—JOHN F. KENNEDY

It's by action that people

have to come to learn what you stand for,

what you're going to do.

—ROBERT F. KENNEDY

We choose to go to the moon.

We choose to go to the moon in this decade

and do the other things,

not because they are easy, but because they are hard,

because that goal will serve to organize

and measure the best of our energies and skills,

because that challenge is one that we are willing to accept,

one we are unwilling to postpone,

and one which we intend to win.

—JOHN F. KENNEDY

Things do not happen. Things are made to happen.

—JOHN F. KENNEDY

The best strategies are always accidental.

—JOHN F. KENNEDY

Now the trumpet summons us again—not as a call

to bear arms, though arms we need;

not as a call to battle, though embattled we are;

but a call to bear the burden of a long twilight struggle,

year in and year out, rejoicing in hope, patient in tribulation,

a struggle against the common enemies

of man: tyranny, poverty, disease, and war itself.

—JOHN F. KENNEDY

So let us not rest all our hopes on parchment and paper,

let us strive to build peace, a desire for peace,

a willingness to work for peace in the hearts and minds of

all our people. I believe that we can. I believe the problems

of human destiny are not beyond the reach of human beings.

—JOHN F. KENNEDY

We can only reach our goal

by the gradual acceptance of the view that we can all gain

more by agreement than by aggression.

—JOHN F. KENNEDY

It takes two to make peace.

—JOHN F. KENNEDY

Peace is a daily, a weekly, a monthly process,

gradually changing directions,

slowly eroding old barriers, quietly building new structures.

—JOHN F. KENNEDY

Men of goodwill, working together,

can grasp the future and mold it to our will.

—ROBERT F. KENNEDY

It is not enough to understand,

or to see clearly. The future will be shaped

in the arena of human activity,

by those willing to commit

their minds and their bodies to the task.

—ROBERT F. KENNEDY

The time to repair the roof is when the sun is shining.

—JOHN F. KENNEDY

As they say on Cape Cod, a rising tide

lifts all boats. And a partnership, by definition, serves both

partners, without domination or unfair advantage.

—JOHN F. KENNEDY

Let me say a few words to all those

that I have met and to all those who have supported me,

at this convention and across the country.

There were hard hours on our journey, and often we sailed

against the wind. But always we kept

our rudder true, and there were so many of you who stayed

the course and shared our hope.

You gave your help, but even more, you gave your hearts.

—EDWARD M. KENNEDY

I'm loyal to the Democratic Party,

but I feel stronger about

the United States and mankind generally.

—ROBERT F. KENNEDY

Let this be our commitment:

Whatever sacrifices must be made will be shared

and shared fairly. And let this be

our confidence: At the end of our journey

and always before us

shines that ideal of liberty and justice for all.

—EDWARD M. KENNEDY

It is the people who must hold

their leaders accountable…. Leadership in turn

must listen and respond.

When both political parties live up to the nation's

truest ideals, we will stand united

in America, and America will be back on track.

—EDWARD M. KENNEDY

What is important is that we

preserve confidence in the good faith of each other....

We will undoubtedly disagree

from time to time on tactics. But the important thing

is to keep in touch.

—JOHN F. KENNEDY

If you're interested in being effective, it's important

to be able to build coalitions. You have to compromise to make

progress. The bottom line is getting things done.

—EDWARD M. KENNEDY

Those who make peaceful revolution

impossible will make violent revolution inevitable.

—JOHN F. KENNEDY

What really is our purpose in life?

For all the advantages we have,

don't we have a major responsibility and

an obligation to those who do not have those advantages?

Don't we have a major responsibility?

—ROBERT F. KENNEDY

Speaking to his children:

In Mississippi, a whole family lives in a shack

the size of this room. The children

are covered with sores and their tummies stick out

because they have no food. Do you know

how lucky you are? Do something for your country.

—ROBERT F. KENNEDY

Writing to his father:

Millions of people need help, Dad. My God, they need help!

I'll make changes. Like Jack.

—ROBERT F. KENNEDY

Governments can err, presidents do make mistakes,

but the immortal Dante tells us

that divine justice weighs the sins of the cold-blooded

and the sins of the warm-blooded

in different scales. Better the occasional faults of a party

living in the spirit of charity

than the consistent omissions of a party frozen

in the ice of its own indifference.

—JOHN F. KENNEDY

Unlike my brothers,

I have been given length of years and time.

And as I approach my sixtieth birthday I am determined

to give all I have to advance the causes

for which I have stood for almost a third of a century.

—EDWARD M. KENNEDY

It is more important to be of service than successful.

—ROBERT F. KENNEDY

What other reason do we

have really for [our] existence as human beings

unless we've made some other contribution

to somebody else to improve their own lives?

—ROBERT F. KENNEDY

PREVIOUS PAGE: President John F. Kennedy greets his son, John Jr., 1963.

LEFT: John F. Kennedy marries Jacqueline Bouvier, Newport, Rhode Island, 1953.

ABOVE: John F. Kennedy and Caroline share a kiss as Jacqueline looks on, Washington, D.C., 1959.

NEXT PAGE: Robert F. and Ethel Kennedy playing and reading with their children before putting them to bed, 1962.

TOP: Edward M. Kennedy with sons, Patrick and Edward Jr., on his sailboat *Mya*, Cape Cod, Massachusetts, 1993.

ABOVE: Senator Kennedy with Caroline and John Jr. at a JFK Profiles in Courage Award ceremony, Boston, Massachusetts, 1995.

RIGHT: Senator Kennedy walks with his young family along the beach, Hyannis Port, Massachusetts, 1969.

NEXT PAGE: Robert F. Kennedy driving his car with his dog Freckles in his lap; his son Max sits beside him, 1966.

Progress is a nice word.

But change is its motivator.

And change has its enemies.

—ROBERT F. KENNEDY

Change is the law of life.

—EDWARD M. KENNEDY

Everything changes but change itself.

—JOHN F. KENNEDY

Sheer violence

cannot compel fundamental change.

—EDWARD M. KENNEDY

I want us to find out

the promise of the future, what we can accomplish

here in the United States,

what this country does stand for and

what is expected of us

in the years ahead. And I also want us to know

and examine where

we've gone wrong. And I want all of us,

young and old, to have a chance

to build a better country and change the direction

of the United States of America.

—ROBERT F. KENNEDY

It is not easy to plant trees

when we will not live to see their flowering.

But that way lies greatness.

And in search of greatness we will find it—for

ourselves as a nation and a people.

—ROBERT F. KENNEDY

[Moral courage] is the one essential,

vital quality for those who seek

to change a world that yields most painfully

to change. And I believe that

in this generation those with the courage

to enter the moral conflict will find themselves with

companions in every corner of the globe.

—EDWARD M. KENNEDY

In such a fantastic and

dangerous world—we will not find

answers in old dogmas,

by repeating outworn slogans,

or fighting on ancient battlegrounds

against fading enemies

long after the real struggle has moved on.

We ourselves must change

to master change. We must rethink all

our old ideas and beliefs

before they capture and destroy us.

And for those answers

America must look to its young people,

the children of this time of change.

—ROBERT F. KENNEDY

The objective of the discontented

should not be revenge,

but change.

—EDWARD M. KENNEDY

Things cannot

be forced from the top.

—JOHN F. KENNEDY

Throughout our history, America has been

blessed that men and women

of conscience, ability, and vision have responded

to the nation's call in times of need.

I think of Rosa Parks, who through her simple,

brave, and eloquent act shamed

a nation into finally confronting the vast discrimination

that many chose to ignore for so long.

Without her act of civil disobedience, how much longer might

we have waited for the spark

of conscience to ignite the determination of millions of

others to call for change? We did not make

the world we live in, but we have the chance to change it.

—EDWARD M. KENNEDY

There is a new wave

of change all around us and if we

set our compass true,

we will reach our destination,

not merely victory

for our party but renewal for our nation.

—EDWARD M. KENNEDY

We have the power to make this

the best generation of mankind in the history of the world—

or to make it the last.

—JOHN F. KENNEDY

On his athletic ability:

I dropped everything. I always fell down.

I always bumped my nose or head.

—ROBERT F. KENNEDY

Our growing softness,

our increasing lack of physical fitness,

is a menace to our security.

—JOHN F. KENNEDY

We are underexercised as a nation. We look

instead of play. We ride instead of walk.

Our existence deprives us of the minimum of

physical activity essential for healthy living.

—JOHN F. KENNEDY

Physical fitness is not only one of

the most important keys to a healthy body, it is the basis of

dynamic and creative intellectual activity.

—JOHN F. KENNEDY

Except for war, there is

nothing in American life—nothing—which

trains a boy better for life than football.

—ROBERT F. KENNEDY

Declining a possible offer to join

the Green Bay Packers after he graduated from Harvard:

[I have plans to] go

into another contact sport—politics.

—EDWARD M. KENNEDY

Number one,

I'm really here not as a presidential candidate,

but to see if Notre Dame will

ever have the courage to play my college, Harvard.

—ROBERT F. KENNEDY

There are not so many differences

between politics and football.

Some Republicans have been unkind enough

to suggest that my close election

was somewhat similar to the Notre Dame-Syracuse game

[won by Notre Dame with a disputed penalty].

But I am like Notre Dame. We just take it as it comes along.

We're not giving it back.

—JOHN F. KENNEDY

I am the man who

accompanied Jacqueline Kennedy to Paris,

and I have enjoyed it.

—JOHN F. KENNEDY

Mrs. Kennedy is organizing herself.

It takes long, but, of course,

she looks better than we do when she does it.

—JOHN F. KENNEDY

To Frank Sinatra about Kennedy brother-in-law

Peter Lawford:

I like the way Peter dresses, Frank, even though

my life's ambition is to be like you.

—JOHN F. KENNEDY

On his brother's Brooks Brothers

buttoned-down shirts:

Bobby doesn't know any better.

—JOHN F. KENNEDY

Jack told me to go upstairs

and comb my hair, to which I said

it was the first time

the president had ever told the attorney general

to comb his hair

before they made an announcement.

—ROBERT F. KENNEDY

Writing to his brother Robert when he was stationed in

the South Pacific during World War II:

The folks sent me a clipping of you taking the oath.

The sight of you there, just as a boy,

was really moving particularly as a close examination

showed that you had my checked London coat on.

I'd like to know what the hell

I'm doing out here while you go stroking around

in my drape coat, but I suppose that [is]

what we are out here for.... In that picture

you looked as if you were going to

step outside the room, grab your gun, and

knock off several of the houseboys before lunch.

—JOHN F. KENNEDY

PART II

PRIVATE LIVES AND FAMILY

PRIVATE LIVES

You've got to live every day

like it is your last day on earth—and

it damn well may be!

—JOHN F. KENNEDY

It is not easy, in the middle of one's life

or political career, to say

that the old horizons are too limited—that

our education must begin again....

The best responses will not be easily found;

nor once found, will they

command unanimous agreement. But

the possibilities of greatness are equal

to the difficulty of the challenge.

—ROBERT F. KENNEDY

We must never forget

that the highest appreciation

is not to utter words,

but to live by them.

—JOHN F. KENNEDY

All of us might wish at times

that we lived in a more tranquil world,

but we don't.

And if our times are difficult

and perplexing,

so are they challenging

and filled with opportunity.

—ROBERT F. KENNEDY

I believe that each of us

as individuals must not only struggle

to make a better world,

but to make ourselves better, too,

and in this life

those endeavors are never finished.

—EDWARD M. KENNEDY

With a good conscience our only sure reward,

with history the final judge of our deeds,

let us go forth to lead the land we love asking His blessing

and His help, but knowing

that here on earth God's work must truly be our own.

—JOHN F. KENNEDY

The future is not

a gift: it is an achievement.

Every generation helps

make its own future.

This is the essential

challenge of the present.

—ROBERT F. KENNEDY

Do not pray

for easy lives. Pray to be

stronger men.

—JOHN F. KENNEDY

Every area of trouble

gives out a ray of hope, and the one

unchangeable certainty

is that nothing

is certain or unchangeable.

—JOHN F. KENNEDY

Hope is one of the qualities

of spirit that make us human. Hope allows us

to dream of a better life

for our children, our community, and our world.

—EDWARD M. KENNEDY

Good luck is something you make,

and bad luck is something you endure.

—ROBERT F. KENNEDY

Tragedy is a tool for the living

to gain wisdom,

not a guide by which to live.

—ROBERT F. KENNEDY

At least half the days that [John Kennedy] spent

on this earth were days of intense physical pain.

I never heard him complain. I never heard him say anything

that would indicate that he felt that God had dealt

with him unjustly. Those who knew him well would know

he was suffering only because his face

was a little white or the lines were a little deeper,

his words a little sharper.

Those who did not know him well detected nothing.

—ROBERT F. KENNEDY

There is a rhythm

to personal and national and international life,

and it ebbs and flows.

—JOHN F. KENNEDY

The fact of the matter is, I have been impacted

over the course of my life by a series of crises,

by a series of tragedies…and I have responded to those

challenges by one, acting responsibly, and two,

by the continuing commitment to public service.

—EDWARD M. KENNEDY

Eventually, in one's soul, you recognize that it's

important to move on. That life's a continuing and unfolding

drama and that you get on with your own life.

—EDWARD M. KENNEDY

Efforts and courage

are not enough

without purpose and direction.

—JOHN F. KENNEDY

I've got to go at a thing

four times as hard and four times as long

as some other fellow.

—EDWARD M. KENNEDY

You've got to learn

to fight your own battles.

—ROBERT F. KENNEDY

But if there was one thing that President Kennedy

stood for...it was the belief that idealism,

high aspiration and deep convictions are not incompatible

with the most practical and efficient of programs—that

there is no basic inconsistency between ideals and realistic

possibilities—no separation between the deepest desires

of heart and of mind and the rational

application of human effort to human problems.

—ROBERT F. KENNEDY

Many years ago, the great British explorer George Mallory,

who was to die on Mount Everest, was asked

why did he want to climb it. He said, "Because it is there."

—JOHN F. KENNEDY

If this capsule history of our progress

teaches us anything, it is that man,

in his quest for knowledge and progress,

is determined and cannot be deterred.

—JOHN F. KENNEDY

It's easy to be discouraged, to be fearful

of the future, to find excuses to justify apathy or indifference

in the face of opposition that seems too strong

or issues that seem too complex. These impulses are

only natural. But you can no more yield

to them than you could walk off the field in the seventh

inning because your team seemed too far behind.

—EDWARD M. KENNEDY

If I hadn't been born rich,

I'd probably be a revolutionary.

—ROBERT F. KENNEDY

All I ask is that I be judged upon my own ability.

—EDWARD M. KENNEDY

I'm going to have a very rough road

ahead of me.... I don't think I'm asking for a free ride.

I don't think I'm asking for a free road.

—ROBERT F. KENNEDY

Speaking to the Civil Rights Commission:

You're second-guessers.

I'm the one who has to get the job done.

—ROBERT F. KENNEDY

I have to do what feels

natural to me. I can't be a hypocrite anymore.

—ROBERT F. KENNEDY

I don't think people would expect me to sit

on my hands for the rest of my life because my brother

is president and my other brother

is attorney general. I wasn't brought up that way.

—EDWARD M. KENNEDY

Responding to what he regretted most:

I wish I had had more good times.

—JOHN F. KENNEDY

There was no such thing as half-trying.

Whether it was running a race

or catching a football, competing in school—we were to try.

And we were to try harder than anyone else.

—ROBERT F. KENNEDY

Intent on learning how to swim

or drowning in the attempt, three-year-old

Robert Kennedy repeatedly threw himself

into the waters of Nantucket Sound.

Recounting the story later, John Kennedy said:

"It either showed a lot of guts, or no sense at all,

depending on how you looked at it."

The only alternative is to give up,

to admit that you are beaten. I have never

admitted that I am beaten.

—ROBERT F. KENNEDY

The pressure of all of the others

on Teddy came to bear so that

he had to do his best. It was a chain reaction

started by Joe, that touched me,

and all my brothers and sisters.

—JOHN F. KENNEDY

Naming his personal strengths:

Perseverance.

—EDWARD M. KENNEDY

To a staffer who wanted to quit:

You're lucky your brother

isn't president of the United States.

—ROBERT F. KENNEDY

To a colleague after noticing

that Jimmy Hoffa's office lights were still on at 1:00 A.M.:

If he's still at work, we ought to be.

—ROBERT F. KENNEDY

Explaining why he always kept calling on a reporter who

continually wrote negative stories about him:

I always say to myself I won't call on her. But she gets up

every time and waves her hand so frantically that toward

the end I look down and she's the only one I seem to see.

—JOHN F. KENNEDY

We soon learned

that competition in the family

was a kind of dry run

for the world outside.

—JOHN F. KENNEDY

I learned a long time ago

that unless you know how to lose,

you don't deserve to win.

—EDWARD M. KENNEDY

Once you say you're going

to settle for second,

that's what happens to you in life.

—JOHN F. KENNEDY

I'm not one of those who think

coming in second or third is winning.

—ROBERT F. KENNEDY

In a letter to an Associated Press reporter

who suggested that he aim no higher than vice president:

I wish you'd get those words out

of your typewriter because

I'm never going to take second place.

—JOHN F. KENNEDY

We want to be first;

not first if, not first but; but first!

—JOHN F. KENNEDY

Failure has no friends.

—JOHN F. KENNEDY

There's an old saying

that victory has one hundred fathers

and defeat is an orphan.

—JOHN F. KENNEDY

Past error is no excuse for its own perpetration…

We do ourselves best justice

when we measure ourselves against ancient texts,

as in Sophocles: "All men make mistakes,

but a good man yields when he knows his course is wrong,

and he repairs the evil."

—ROBERT F. KENNEDY

On being caught cheating while he was a student at Harvard:

What I did was wrong

and I have regretted it ever since. The unhappiness

I caused my family and friends,

even though eleven years ago, has been a bitter experience

for me, but it has also been a very valuable lesson.

—EDWARD M. KENNEDY

Well, I've made mistakes in my past,

but I've always tried to learn from them.

And I've always tried, every day, to be a better person.

And being down here reminds me

of all the forces in my life that were just so powerful

and continue to be inspirational to me,

the great sense of family that I have being here,

and they remind me of those important forces,

the importance of faith in my life.

—EDWARD M. KENNEDY

After reading President Eisenhower's memoirs:

Apparently Ike never did anything wrong....

When we come to writing

the memoirs of this administration,

we'll do it differently.

—JOHN F. KENNEDY

When the history is going to be written

about this conflict [the Vietnam War], I'm obviously going to

have to take my share of personal responsibility.

I happen to think I learned something from that.

—ROBERT F. KENNEDY

Let us admit that as all men make mistakes, so do nations...

—EDWARD M. KENNEDY

It isn't that I'm a saint.

It's just that I've never found it

necessary to be a sinner.

—ROBERT F. KENNEDY

He's a puritan, absolutely incorruptible.

Then he had terrific executive energy.

We've got more guys around here with ideas.

The problem is to get things done.

Bobby's the best organizer I've ever seen.

—JOHN F. KENNEDY

Let's face it.

I appeal best to people who have problems.

—ROBERT F. KENNEDY

I suppose if you had to choose

just one quality to have that would be it: vitality.

—JOHN F. KENNEDY

I'm an idealist without illusions.

—JOHN F. KENNEDY

It doesn't matter

if they like me or not. Jack can be nice to them...

Somebody has to be able to say no.

—ROBERT F. KENNEDY

I've learned you can't rely on altruism or morality.

People just aren't built that way.

—ROBERT F. KENNEDY

How he hoped to be remembered:

As the guy who broke the Mafia.

—ROBERT F. KENNEDY

I feel a special obligation to those who share my hopes for

this state and nation, who in the past have given me

their help—and often even their hearts… I am painfully aware

that the criticism directed at me in recent years involves far

more than honest disagreement with my positions…

It also involves the disappointment of friends and many others

who rely on me to fight the good fight. To them I say,

I recognize my own shortcomings, the faults in the conduct

of my private life. I realize that I alone am

responsible for them. I am the one who must confront them.

—EDWARD M. KENNEDY

[The voters] need to see me,

to be convinced I'm reliable and mature.

You can't counter

the Chappaquiddick thing directly.

The answer has to be

implicit in who you are, what you stand for

and how they see you.

—EDWARD M. KENNEDY

The votes you cast for me tonight

were not just for a senator, but for a cause....

Looking out across this room tonight,

I see great friends over many years who were there

for Jack in 1960.... I see supporters and volunteers

and foot soldiers from my own first campaign in 1962,

and many of Bobby's supporters as well.

You've been there with us through good times

and bad, and I'll never forget you....

The poet Yeats said it well—

"Think where man's glory most begins and ends,

and say my glory was I had such friends."

—EDWARD M. KENNEDY

The presidency

is not a very good place

to make new friends.

—JOHN F. KENNEDY

You won't have

any trouble finding my enemies.

They're all over town.

—ROBERT F. KENNEDY

My first love is Jimmy Hoffa.

—ROBERT F. KENNEDY

What it really all adds up to is love—

not love as it is described with such facility

in popular magazines, but the kind

of love that is affection

and respect, order, encouragement,

and support. Our awareness

of this was an incalculable source of strength,

and because real love is

something unselfish and involves

sacrifice and giving, we could not help

but profit from it.

—ROBERT F. KENNEDY

I believe deeply in hope.

I believe,

you know, in the power of love.

—EDWARD M. KENNEDY

Don't tell me

people in this country

don't love me.

—ROBERT F. KENNEDY

I love [Jackie] deeply

and have done everything for her.

I've no feeling of letting her down

because I've put her foremost in everything.

—JOHN F. KENNEDY

Happiness:

The full use of your powers

along lines of excellence.

—JOHN F. KENNEDY

Asked when and where he was happiest:

Every Thanksgiving,

when our family gathers on Cape Cod.

—EDWARD M. KENNEDY

There are three things in life

which are real: God, human folly, and laughter.

Since the first two are beyond

our comprehensions,

we must do what we can with the third.

—JOHN F. KENNEDY

[My mother] sustained us

in the saddest times by her faith in God,

which was the greatest gift she gave us....

She was ambitious not only for our success

but for our souls.

—EDWARD M. KENNEDY

[Mother] will be there ready

to welcome the rest of us home someday,

of this I have no doubt,

as they were from the very beginning,

Mother's prayers will continue

to be more than enough to bring us through.

—EDWARD M. KENNEDY

The lessons we drew from our religious faith

influence our values and our vision

of what America could be. For me, the most profound message

is in the Gospel of Matthew:

"For I was hungry and you gave me food, I was thirsty

and you gave me drink, a stranger

and you welcomed me, naked and you clothed me, ill and

you cared for me, in prison and you visited me."

—EDWARD M. KENNEDY

After speaking at the Al Smith dinner during

the 1960 presidential election and noticing that a lot

of the audience—many of whom were wealthy Catholics—

seemed to be pro-Nixon, John Kennedy remarked:

It all goes to show that when the chips are down,

money counts more than religion.

After the death of his newborn son Patrick, in 1963:

It is against the laws of nature

for parents to bury their children.

—JOHN F. KENNEDY

Telling John F. Kennedy Jr. that his father had died:

Your daddy has gone to heaven to be with Patrick.

Patrick was very lonely. He didn't know anybody in heaven.

Now he is being taken care of by a great friend.

—ROBERT F. KENNEDY

On his brother's death:

Every time I look at [Caroline],

I want to go somewhere and cry.

—ROBERT F. KENNEDY

When there were difficulties, you sustained him.

When there were periods of crisis, you

stood beside him. When there were periods of happiness,

you laughed with him. And when there were

periods of sorrow, you comforted him. I realize that

as individuals we can't just look back, that we must

look forward. When I think of President Kennedy,

I think of what Shakespeare said in

Romeo and Juliet: "When he shall die take him

and cut him out into stars and he shall make

the face of heaven so fine that all the world will be in love

with night and pay no worship to the garish sun."

—ROBERT F. KENNEDY

After being elected senator:

Well, if my brother were alive, I wouldn't be here.

I'd rather have it that way.

—ROBERT F. KENNEDY

Asked what he regards as the depth of misery:

Suddenly losing a loved one.

—EDWARD M. KENNEDY

On the family members who have died:

A day doesn't go by where I'm not thoughtful about them

and don't miss them. I mean, I do in a very, very real way.

And it's still a very—raw occasion—

they're very close to the surface in terms of my life and,

I think, for all of the members of our family.

—EDWARD M. KENNEDY

If someone is going

to kill me, they are going to kill me.

—JOHN F. KENNEDY

Living each day

is like Russian roulette.

—ROBERT F. KENNEDY

Upon receiving a letter from Nikita Khrushchev,

the Soviet premier, during the Cuban Missile Crisis,

John Kennedy said:

"This is the night to go to the theater,

just like Abraham Lincoln."

Robert Kennedy responded:

"If you go, I want to go with you."

I thought they'd get one of us, but Jack,

after all he's been through, never worried about it.

I thought it would be me.

—ROBERT F. KENNEDY

On the Kennedy curse:

I guess the only reason we've survived

is that there are more

of us than there is trouble.

—ROBERT F. KENNEDY

I don't know that it makes any difference

what I do. Maybe we're all doomed anyway.

—ROBERT F. KENNEDY

[T]o those nations who would make

themselves our adversary,

we offer not a pledge but a request—that both sides

begin anew the quest for peace,

before the dark powers of destruction

unleashed by science

engulf all humanity in planned

or accidental self-destruction.

—JOHN F. KENNEDY

Circumstances may change,

but the work

of compassion must continue.

—EDWARD M. KENNEDY

To a crowd who had just found out

about Martin Luther King Jr.'s assassination:

What we need in the United States is not division;

what we need in the United States is not hatred;

what we need in the United States is not violence and lawlessness,

but is love and wisdom, and compassion toward one another,

and a feeling of justice toward those who still suffer

within our country, whether they be white

or whether they be black.

—ROBERT F. KENNEDY

On his opposition

to the death sentence for Sirhan Sirhan,

Robert Kennedy's assassin:

My brother was a man of love and sentiment

and compassion.

He would not have wanted his death

to be a cause for

the taking of another life.

—EDWARD M. KENNEDY

After all, in the New Testament,

even the disciples had to be taught to look first

to the beam in their own eyes,

and only then to the mote in their neighbor's eyes.

—EDWARD M. KENNEDY

To students in Derry, Northern Ireland:

Like so many of you here,

my family has been touched by tragedy.

I know that the feelings

of grief and loss are immediate—and

they are enduring.

The best way to ease these feelings

is to forgive, and to carry on—

not to lash out in fury,

but to reach out in trust and hope.

—EDWARD M. KENNEDY

FAMILY

My brother was really smitten with [Jackie]

right from the beginning.... He was fascinated

by her intelligence: they read together,

painted together, enjoyed good conversation

and walks together.

—EDWARD M. KENNEDY

On being asked to go out for drinks with a friend:

I'm sorry…I got engaged to be married two weeks ago.

—JOHN F. KENNEDY

On the secret of John and Jacqueline Kennedy's marriage:

Jack knows she'll never greet him with, "What's new in Laos?"

—ROBERT F. KENNEDY

Whenever a wife says anything in this town, everyone

assumes that she is saying what her husband really thinks.

Imagine how I felt last night when I thought I heard

Jackie telling Malraux that Adenauer was "*un peu gaga*!"

—JOHN F. KENNEDY

Asked if he married his second wife, Vicki, for political reasons:

Vicki knows how much I love her.

I have a sense about how much she loves me.

My children know what it means for us to be together....

And I think the people that see us together

are coming to know that. And that's good enough for me.

—EDWARD M. KENNEDY

We rely on our youth

for all our hopes of a better future...for

the very meaning of our lives.

—ROBERT F. KENNEDY

A child miseducated is a child lost.

—JOHN F. KENNEDY

In the last analysis the quality of education

is a question of commitment—of whether people like us

are willing to go into the classrooms

as teachers or parents, as volunteers or just as

concerned citizens, to ensure that every child learns

to the full limit his capabilities.

—ROBERT F. KENNEDY

Without a decent home, without an environment of equality,

our hopes for our children are merely illusions.

—ROBERT F. KENNEDY

Each generation makes its own accounting according to its children.

—ROBERT F. KENNEDY

Children are the world's most valuable

resource and its best hope for the future.

—JOHN F. KENNEDY

Finding his daughter Caroline standing outside the Cabinet Room:

Have you been eating candy?

Caroline, have you been eating candy? Answer me. Yes, no, or maybe.

—JOHN F. KENNEDY

On the Cuban Missile Crisis:

If it weren't for the children,

it would be so easy to press the button.

Not just John and Caroline,

and not just the children in America,

but children all over the world

who will suffer and die

for the decision I have to make.

—JOHN F. KENNEDY

It is our task in our time and in our generation

to hand down undiminished

to those who come after us, as was handed down

to us by those who went before,

the natural wealth and beauty which is ours.

—JOHN F. KENNEDY

I was the seventh of nine children, and when you come

from that far down you have to struggle to survive.

—ROBERT F. KENNEDY

On his greatest accomplishment:

Having children who turned out to be the living,

involved, and interesting people they are.

—EDWARD M. KENNEDY

On the difficulties of caring for his son, Edward Kennedy Jr.,

who had bone cancer when he was twelve:

He's a courageous boy, but it is very tough…. The silver streak

in that grim and difficult time is that it has formed a bond

between him and me that has just been…extraordinary.

I can't tell whether it would have been so or not otherwise.

—EDWARD M. KENNEDY

PREVIOUS PAGE: Presidential candidate John F. Kennedy and Edward M. Kennedy confer during a campaign stop in West Virginia, 1960.

ABOVE: Robert F. Kennedy campaigning in Crawfordsville, Indiana, 1968.

OPPOSITE: Kennedy waves to supporters while campaigning for the Democratic presidential nomination, Detroit, Michigan, 1968.

NEXT PAGE: Edward M. Kennedy speaks at the Vietnam Veterans Against the War demonstration, Washington, D.C., 1972. Future Massachusetts senator John Kerry (top left) sits in the crowd.

TOP: John F. Kennedy reading an official document, 1961; MIDDLE: playing with Caroline and John Jr. in the Oval Office, 1962; BOTTOM: walking outside the White House after Sunday Mass, 1961; RIGHT: resting in his Boston apartment, ca. 1959–1960.

NEXT PAGE: Edward M. Kennedy with Barack Obama at George W. Bush's final State of the Union Address, Washington, D.C., 2008. That same day, Kennedy endorsed Obama as the Democratic presidential candidate.

My father wasn't around as much as some fathers

when I was young; but whether he was

there or not, he made his children feel that they were

the most important things in the world to him.

He was so terribly interested in everything we were doing.

He held up standards for us, and he was very tough

when we failed to meet those standards.

The toughness was important.

If it hadn't been for that, Teddy might be just

a playboy today. But my father

cracked down on him at a crucial time in his life,

and this brought out in Teddy

the discipline and seriousness which will

make him an important political figure.

—JOHN F. KENNEDY

I always felt the greatest gift that Dad gave to each of us,

was his unqualified support of any and all of our

undertakings.... He was also quick to admonish us for errors.

He tolerated a mistake once, but never a second time.

—EDWARD M. KENNEDY

[My father] loved all of us—the boys in a very special way.

I can say that, except for his influence and encouragement,

my brother Jack might not have run for the Senate in 1952,

there would have been much less likelihood

that he would have received the presidential nomination in 1960,

I could not have become attorney general,

and my brother Teddy would not have run for the Senate in 1962.

—ROBERT F. KENNEDY

The great thing about Dad

is his optimism and his enthusiasm and

how he's always for you.

He might not always agree with what I do,

just as I don't always agree with him,

but as soon as I do anything, there's Dad saying,

"Smartest move you ever made....

We really got them on the run now."

—JOHN F. KENNEDY

I guess Dad has decided

that he's going to be a ventriloquist, so I guess

that leaves me in the role of dummy.

—JOHN F. KENNEDY

Henry the Fourth. That's my Dad.

—ROBERT F. KENNEDY

The fact [my mother is] one hundred

has not diminished the power of her presence,

the source of inspiration,

the fact she continues to be the central

figure of our family,

the safe harbor in a storm.

And all of her children or grandchildren

or great-grandchildren

continue to be inspired by her.

—EDWARD M. KENNEDY

As we gathered to share memories of Mother,

grandchild after grandchild

stood to tell anecdotes about Mother—different stories

with one common theme.

She instilled in the next generation the bonds of faith

and love that tie us together as a family.

—EDWARD M. KENNEDY

Mother always thought her children

should strive for high places.

But inside the family, with love and laughter,

she knew how to put each of us in our place.

—EDWARD M. KENNEDY

Seeing Caroline ride Macaroni, her pet horse,

through the Rose Garden:

Well, well, if it isn't Hopalong Cassidy in the flesh.

—JOHN F. KENNEDY

After winning the California primary:

I want to express gratitude to my dog Freckles…

and I'm not doing this is any order of importance,

but I also want to thank my wife Ethel.

—ROBERT F. KENNEDY

If a crowd was small, Freckles would go to sleep.

Robert Kennedy would then summon his aides, saying:

"The dog's getting pretty upset with this crowd."

On bringing Brumus, his Newfoundland,

to the Justice Department:

He usually stays at home with the children.

But the children are away on vacation and he gets very lonely.

So I bring him down here and

get pretty girls to take him for walks.

—ROBERT F. KENNEDY

On Splash, his Portuguese water dog:

He's met Elton John. He was in the Oval Office.

He has a dog bone from President Bush. He gave me this

rawhide dog bone and wrote on it, "From Barney to Splash."

I take him to all the hearings; he always sits under the table.

He goes to press conferences, to the Cape. He loves to take

long trips in the boat. He could sail all day.

—EDWARD M. KENNEDY

How he became a hero:

It was involuntary. They sank my boat.

—JOHN F. KENNEDY

When I came to Washington

to the U.S. Senate, I brought a number of young ladies

from Massachusetts to be secretaries.

They all got married. Then I got a whole new set of girls

and they got married. So if any of you girls

feel the prospects are limited in this community

you come and work for me.

—JOHN F. KENNEDY

Let's not talk so much

about vice. I'm against vice, in all forms.

—JOHN F. KENNEDY

Replying to Arthur Schlesinger's comment that

he "paid a heavy price" for naming his book

Profiles in Courage:

Yes, but I didn't have a chapter in it on myself.

—JOHN F. KENNEDY

On the inaugural balls:

The Johnsons and I have been to five balls tonight,

and we still have one unfulfilled ambition—

and that is to see somebody dance.

—JOHN F. KENNEDY

Inscription inside a leather,

gold-embossed copy of a book

he gave to his brother Robert for Christmas:

To the Brother Within—who made the easy difficult.

—JOHN F. KENNEDY

Explaining why he chose his brother for attorney general:

I don't see what's wrong with giving Bobby

a little experience before he starts to practice law.

—JOHN F. KENNEDY

How he proposed to announce his pick for attorney general:

Well, I think I'll open the front door of the Georgetown house

some morning at two A.M., look up and down the street,

and if there's no one there, I'll whisper, "It's Bobby."

—JOHN F. KENNEDY

People say I am ruthless. I am not ruthless.

And if I find the man who is calling me ruthless, I shall destroy him.

—ROBERT F. KENNEDY

Teddy has been down in Washington, and he came to see me

the other day, and he said he was really tired

of being referred to as the kid brother and just another Kennedy,

so he was going to change his name and go out on his own.

—JOHN F. KENNEDY

Speaking at a dinner honoring Nobel Prize winners:

I think this is the most extraordinary collection of talent,

of human knowledge, that has ever been gathered together

at the White House—with the possible exception

of when Thomas Jefferson dined alone.

—JOHN F. KENNEDY

When a photographer said to Robert Kennedy,

"Step back a little, you're casting

a shadow on Ted," Edward Kennedy wisecracked:

"It's going to be the same in Washington."

When asked if he had anything in mind

if he didn't run for president, Edward Kennedy responded:

"Maybe return to law practice in Boston.

Maybe see if I can buy the *Boston Herald*.

Maybe sit in the South of France."

Asked what he would come back as after his death:

Probably a punching bag.

—EDWARD M. KENNEDY

If you wanted something special,

you had to make a pretty good case for it.

—EDWARD M. KENNEDY

I can hardly remember a mealtime

when the conversation

was not dominated by what Franklin D. Roosevelt

was doing or what

was happening around the world....

Since public affairs

had dominated so much of our actions

and discussions, public life

seemed really an extension of family life.

—ROBERT F. KENNEDY

———◦❖◦———

One of my more vivid childhood memories

is of our family gatherings

around the table at dinnertime…. We learned early

that the way to be an active part

of dinner conversation was to have read a book,

to have learned something new in school,

or, as we got older, to have traveled to new places.

—EDWARD M. KENNEDY

In our family, nobody briefed anyone on

how to conduct himself in a job. We just picked things up by

watching and absorbing. That's one advantage

of being a Kennedy—there were so many of us, doing so much.

You just soaked things up as you went along.

—EDWARD M. KENNEDY

———◦❖◦———

To his brother Edward:

You're a Kennedy. Take care of yourself.

—ROBERT F. KENNEDY

In a letter to his son Joe:

You are the oldest of all the male grandchildren.

You have a special and particular

responsibility now which I know you will fulfill. Remember

all the things that Jack started—be kind to others that are

less fortunate than we—and love our country.

—ROBERT F. KENNEDY

To his children:

Kennedys never cry. Kennedys never give up.

—ROBERT F. KENNEDY

Whatever contributions the Kennedys

have made are very much tied

into the incredible importance and power

of that force in our lives—the family.

—EDWARD M. KENNEDY

Writing to his mother:

The trouble with Thanksgiving,

is that you get home just long enough

to see how much you are missing.

—JOHN F. KENNEDY

Helping his father stand after his stroke:

That's what I'm here for, Dad.

Just to give you a hand when you need it.

You've done that for me all my life,

so why can't I do the same for you now?

—ROBERT F. KENNEDY

Writing to his father about his brother Joe:

I think that if the Kennedy children

amount to anything now or ever, it will be due

more to Joe's behavior and

his constant example than to any other factor.

—JOHN F. KENNEDY

On Robert Kennedy:

He's the only one

who doesn't stick knives in my back,

the only one I can count on

when it comes down to it.

—JOHN F. KENNEDY

[My brothers] were my heroes,

my best friends,

and they were great sources

of inspiration for me.

So I didn't really have to

reach outside of my family

to find my heroes.

—EDWARD M. KENNEDY

I sort of hung out with them as a younger brother

who was full of...stars in his eyes.

—EDWARD M. KENNEDY

He and I had a special bond,

despite the fourteen years between us. When I was born,

he asked to be my godfather. He was the best man

at my wedding. He taught me to ride a bicycle, to throw

a forward pass, to sail against the wind.... That is

the way it was with Jack. There was a sense of progress and

adventure, a rejection of complacency and conformity.

There was a common mission, a shared ideal, and above all

the joy of high purpose and great achievement.

—EDWARD M. KENNEDY

Bobby was the one who used to call me

up to see how I was getting along.

On the two or three weekends I was able to

get off from the boarding school

outside Boston, he'd spend the weekend with me.

I'll never forget how we used to

go to the big, empty house at Cape Cod—just

the two of us, rattling around alone.

But Bobby was in charge,

taking care of me, and always making sure

I had something to do.

—EDWARD M. KENNEDY

[Robert Kennedy] gave us strength

in time of trouble,

wisdom in time of uncertainty,

and sharing in time of happiness.

He was always by our side.

Love is not an easy feeling to put into words.

Nor is loyalty, or trust or joy.

But he was all of these.

He loved life completely and lived it intensely.

—EDWARD M. KENNEDY

I think about my brothers every day.

—EDWARD M. KENNEDY

All of us of Irish descent

are bound together by the ties that come from

a common experience; experience which may exist only

in memories and in legend

but which is real enough to those who possess it.

—JOHN F. KENNEDY

Speaking in Ireland:

It took a hundred and fifteen years to make this trip, and

six thousand miles, and three generations,

but I am proud to be here. When my great-grandfather

left here to become a copper in East Boston, he carried

nothing with him except two things: a strong religious faith

and a strong desire for liberty. I am glad to say that

all of his great-grandchildren have valued that inheritance.

—JOHN F. KENNEDY

The White House was designed by Hoban,

a noted Irish-American architect,

and I have no doubt that he believed by incorporating

several features of the Dublin style he

would make it more homelike for any president of Irish

descent. It was a long wait, but I appreciate his efforts.

—JOHN F. KENNEDY

I am constantly reminded of my immigrant heritage.

—EDWARD M. KENNEDY

On a clear day, you can see to Ireland [from Hyannis Port].

We should skip a stone for good luck

toward the island of our Kennedy and Fitzgerald ancestors.

—EDWARD M. KENNEDY

BIBLIOGRAPHY

BOOKS

Adler, Bill, ed. *The Kennedy Wit*. New York, NY: Citadel Press, 1964.

Anderson, Christopher. *Jack and Jackie: Portrait of an American Marriage*.
New York, NY: William Morrow and Company, 1996.

Anthony, Carl Sferrazza. *The Kennedy White House: Family Life and Pictures,
1961–1963*. New York, NY: Simon & Schuster, 2001.

Beran, Michael Knox. *The Last Patrician: Bobby Kennedy and the End of an American
Aristocracy*. New York, NY: St. Martin's Press, 1998.

Burns, James McGregor. *Edward Kennedy and the Camelot Legacy*. New York, NY:
W. W. Norton & Company, 1976.

Clarke, Thurston. *The Last Campaign: Robert F. Kennedy and 82 Days That Inspired
America*. New York, NY: Henry Holt and Company, 2008.

Clymer, Adam. *Edward M. Kennedy: A Biography*. New York, NY: William Morrow
and Company, 1999.

Cook, John. *The Book of Positive Quotations*. Minneapolis, MN: Fairview Press, 2007.

David, Lester and Irene David. *Bobby Kennedy: The Making of a Folk Hero*.
New York, NY: Dodd, Mead & Company, 1986.

David, Lester. *Good Ted, Bad Ted: The Two Faces of Edward M. Kennedy*. New York, NY:
Carol Publishing Group, 1993.

Hamilton, Nigel. *JFK: Reckless Youth*. New York, NY: Random House, 1992.

Hersh, Burton. *The Shadow President: Ted Kennedy in Opposition*. South Royalton, VT: Steerforth Press, 1997.

Heymann, David C. *American Legacy: The Story of John & Caroline Kennedy*. New York, NY: Atria Books, 2007.

_____. *RFK: A Candid Biography of Robert F. Kennedy*. New York, NY: Dutton, 1998.

Kennedy, Edward M. *America Back on Track*. New York, NY: Penguin Group, 2007.

Kennedy, John F. *Profiles in Courage: Decisive Moments in the Lives of Celebrated Americans*. New York, NY: HarperCollins Publishers, 1956.

_____. *Prelude to Leadership: The Post-War Diary of John F. Kennedy*. Washington, D.C.: Regnery Publishing, Inc., 1995.

_____. *Quotations of John F. Kennedy*. Bedford, MA: Applewood Books, 2008.

Kennedy, Robert. *In His Own Words*. New York, NY: Bantam Books, 1988.

Klein, Edward. *All Too Human: The Love Story of Jack and Jackie Kennedy*. New York, NY: Pocket Books, 1996.

Leaming, Barbara. *Jack Kennedy: The Education of a Statesman*. New York, NY: W.W. Norton & Company, 2006.

Mahoney, Richard D. *Sons & Brothers*. New York, NY: Arcade Publishing, 1999.

Maier, Thomas. *The Kennedys: America's Emerald Kings*. New York, NY: Basic Books, 2003.

McGinniss, Joe. *The Last Brother*. New York, NY: Simon & Schuster, 1993.

Moody, Sidney C., ed. *Triumph and Tragedy: The Story of the Kennedys*. Washington, D.C.: Associated Press, 1968.

Newfield, Jack. *Robert Kennedy: A Memoir*. New York, NY: Dutton, 1969.

Perret, Geoffrey. *Jack: A Life Like No Other*. New York, NY: Random House, 2001.

Reeves, Thomas C. *A Question of Character: The Life of John F. Kennedy*. New York, NY: The Free Press, 1991.

Rubin, Gretchen. *Forty Ways to Look at JFK*. New York, NY: Ballantine Books, 2005.

Schlesinger, Arthur M. *A Thousand Days: John F. Kennedy in the White House*. New York, NY: Houghton Mifflin, 2002.

_____. *Robert Kennedy and His Times*. New York, NY: Houghton Mifflin, 2002.

Smith, Sally Bedell. *Grace and Power: The Private World of the Kennedy White House*. New York, NY: Random House, 2004.

Steel, Ronald. *In Love with Night: The American Romance with Robert Kennedy*. New York: Simon and Schuster, 2000.

Talbot, David. *The Hidden History of the Kennedy Years*. New York, NY: Free Press, 2007.

Thomas, Evan. *Robert Kennedy: His Life*. New York, NY: Simon & Schuster, 2000.

BIBLIOGRAPHY

ARTICLES

Milligan, Susan. "Making a Splash." *The Boston Globe*, www.boston.com, May 2006.

"Proust Questionnaire: Edward M. Kennedy." *Vanity Fair*, www.vanityfair.com, May 2006.

WEB SITES

American Rhetoric, www.americanrhetoric.com

CNN, www.cnn.com

Edward M. Kennedy, www.tedkennedy.com

John F. Kennedy Library, www.jfklibrary.org

Robert F. Kennedy Memorial, www.rfkmemorial.org

The History Place, www.historyplace.com

YouTube, www.youtube.com

SET YOUR COMPASS TRUE
THE WISDOM OF JOHN, ROBERT & EDWARD KENNEDY

HarperCollins books may be purchased for educational, business, or sales promotional use. For information, please write: Special Markets Department, HarperCollins*Publishers*, 10 East 53rd Street, New York, NY 10022.

First published in 2009 by
Collins Design
An Imprint of HarperCollins*Publishers*
10 East 53rd Street
New York, NY 10022
Tel: (212) 207-7000
Fax: (212) 207-7654
collinsdesign@harpercollins.com
www.harpercollins.com

Distributed throughout the world by
HarperCollins*Publishers*
10 East 53rd Street
New York, NY 10022
Fax: (212) 207-7654

INTERIOR DESIGN BY AGNIESZKA STACHOWICZ
JACKET DESIGN BY ARCHIE FERGUSON

Library of Congress Control Number:
2008941584

ISBN 978-0-06-179279-3

Printed in the United States
First Printing, 2009

For all those whose cares
have been our concern,
the work goes on,
the cause endures, the hope
still lives, and
the dream shall never die.

—EDWARD M. KENNEDY